AGV-1503
14.00 ODC
B-17

526.
Z52e

SPACE MISSIONS™

The *Endeavour* SRTM

Mapping the Earth

Helen Zelon

The Rosen Publishing Group's
PowerKids Press™
New York

For Nathaniel Sasson, partner for life

Published in 2002 by The Rosen Publishing Group, Inc.
29 East 21st Street, New York, NY 10010

Copyright © 2002 by The Rosen Publishing Group, Inc.

All rights reserved. No part of this book may be reproduced in any form without permission in writing from the publisher, except by a reviewer.

First Edition

Book Design: Michael de Guzman
Project Editors: Jennifer Landau, Jason Moring, Jennifer Quasha

Photo Credits: pp. 4, 7, 8, 11, 12, 15 © Photri-Microstock; p.16 courtesy of NASA/JPL/California Institute of Technology; p. 19 © AFP/CORBIS; p. 20 © Reuters New Media Inc./CORBIS.

Zelon, Helen.
 The Endeavour SRTM : mapping the earth / Helen Zelon.— 1st ed.
     p. cm. — (Space missions)
Includes bibliographical references and index.
ISBN 0-8239-5775-6 (lib. bdg.)
1. Cartography—Remote sensing—Juvenile literature. 2. Radar in earth sciences—Juvenile literature. 3. Maps, Topographic—Juvenile literature. 4. Endeavour (Space shuttle)—Juvenile literature.
[1. Endeavour (Space shuttle) 2. Radar in earth sciences. 3. Cartography—Remote sensing.] I. Title.
GA102.4.R44 Z45 2002
 526—dc21

2001000142

Manufactured in the United States of America

Contents

1	Mapping the World	5
2	Working Together	6
3	Radar Vision	9
4	Eyes on Earth	10
5	In Orbit	13
6	Working in Space	14
7	Interview with Kevin Kregel	16
8	Communication	18
9	Building the Map	21
10	Coming Home	22
	Glossary	23
	Index	24
	Web Sites	24

Mapping the World

On Friday, February 11, 2000, the space shuttle *Endeavour* lifted off the **launchpad** at Kennedy Space Center in Florida. The *Endeavour* had six astronauts on board. Commander Kevin Kregel, pilot Dominic Gorie, and **mission specialists** Janet Kavandi and Janice Voss were Americans. The other astronauts were Mamoru Mohri of Japan and Gerhard Thiele of the European Space Agency. The purpose of the Shuttle **Radar Topography** Mission (SRTM) was to map the world's landmasses using radar. Topographic maps show the shape of the land. Even the best of these kinds of maps don't show earthquake **fault lines** and volcanic activity. Topographic maps made using radar would be able to show these things.

← *This picture shows the* Endeavour *taking off from the launchpad at the start of the Shuttle Radar Topography Mission.*

Working Together

Over 11 days and 181 **orbits**, the Shuttle Radar Topography Mission mapped 80 percent of Earth. This map covered landmasses from Saint Petersburg, Russia, to the southern tip of South America. More than 95 percent of Earth's people live in this area. Scientists planned to use information from the SRTM project to see flooding patterns, **landslides**, and earthquakes. Pilots could fly more safely with better maps, and firefighters could reach faraway blazes more easily. Different countries worked together to make the mission a success. Three agencies were American, including **NASA**, the Jet **Propulsion** Laboratory in California, and the National Imagery and Mapping Agency in Washington, D.C. The German and Italian space agencies also helped with the mission.

Top: The highlighted sections on this picture of Earth show some of the areas the SRTM planned to map. Bottom: This image of the Fiji Islands was made using information from the SRTM.

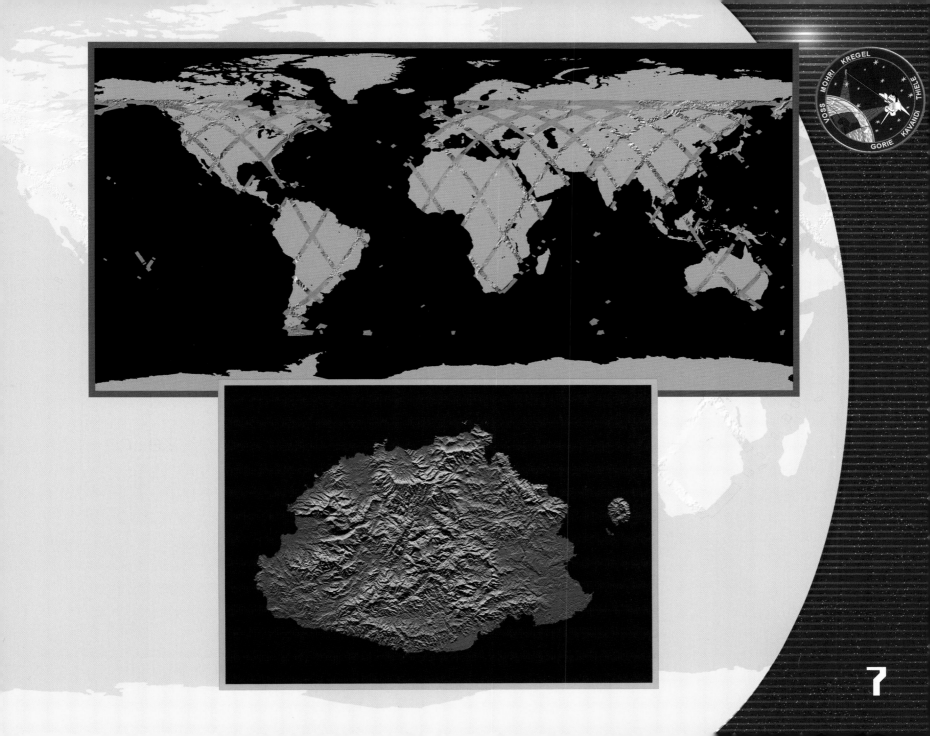

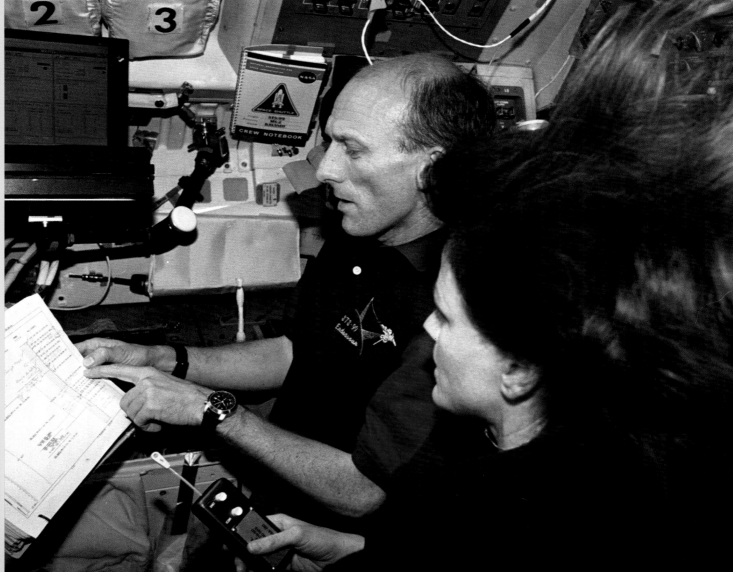

Radar Vision

Traveling on the *Endeavour* space shuttle, the SRTM system mapped the world by sending radar signals from its two **antennae** down to Earth. The signals then bounced back up to the two SRTM antennae. Radar uses sound waves made by a radio to "see" objects. The time it takes the sound wave to reach an object and return to the antenna tells how far away it is. Airline pilots, ship captains, and police officers use radar to track planes, locate ships at sea, and find speeding cars. Radar works as well at night as it does during the day because it uses sound, not light, to see. It also can be used in any weather condition. This makes radar the perfect way to map Earth from space.

◀ *In this picture, astronauts Gerhard Thiele and Janet Kavandi check information coming from the* Endeavour's *two antennae.*

Eyes on Earth

The SRTM used radar to map Earth. The system had two antennae because one antenna would have given only one flat image. In the same way that we use our two eyes to see **depth**, the SRTM used radar from two places on the shuttle. The information from the two radar eyes showed scientists the shapes of Earth's landmasses.

For the SRTM, a special radar unit was built into the space shuttle *Endeavour's* **payload bay**. The second radar system was its mast, a 200-foot-long (61-m-long) pole with a radar unit on its tip. This mast is the longest **rigid** structure ever to fly in space. It began the trip folded up inside a 9-foot (3-m) **canister**. Once the shuttle reached orbit, the canister opened and the mast unfolded to its full length.

The top left and bottom pictures use art and computer-made images to show how the Endeavour *mapped Earth. The top right picture, created by an artist, shows the* Endeavour *starting to unfold its radar mast.*

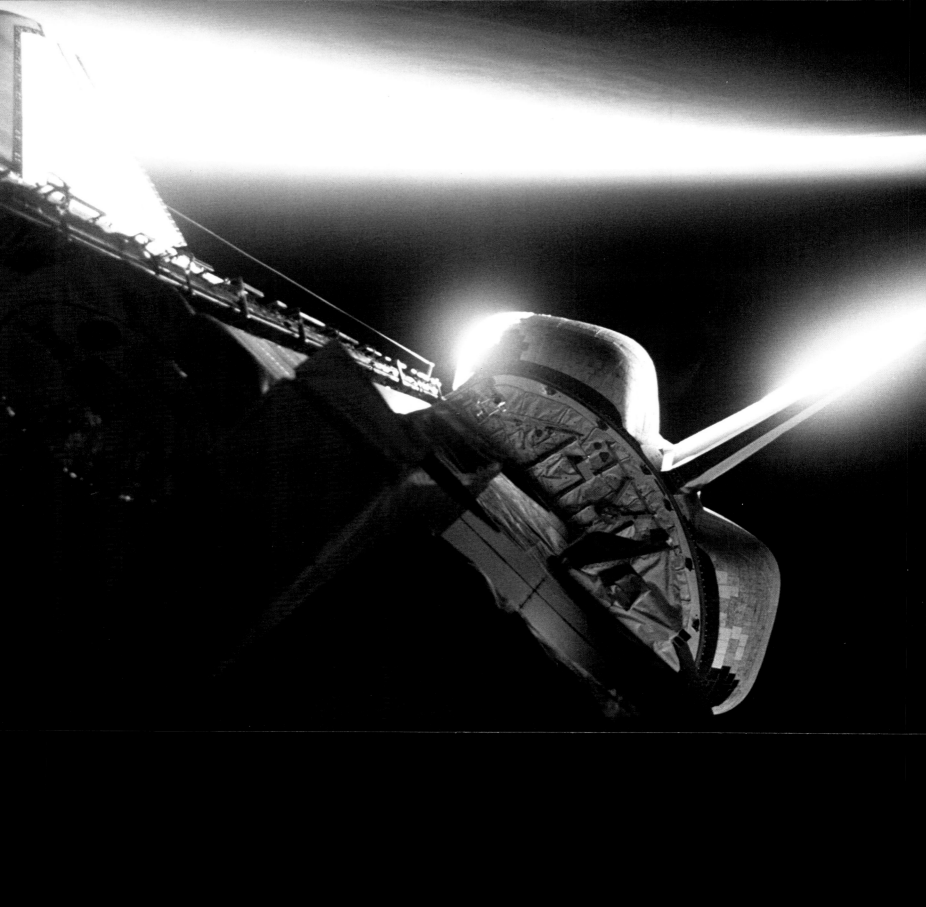

In Orbit

Only 8 minutes after its launch on Friday, February 11, 2000, the *Endeavour* settled into orbit. The space shuttle orbited at 17,000 mph (27,359 km/h), about 150 miles (241 km) above Earth. The astronauts opened the payload bay doors and got ready to unfold the SRTM mast. The astronauts tested the **thruster rockets** to make sure the shuttle's orbital path could be changed, if necessary. Next the astronauts set up the tape recorders that would collect the information the radar system sent back to the shuttle. The SRTM mast stretched outside the payload bay like a long arm. It worked perfectly. By lunch the next day, the radar system had mapped 17 million square miles (44 million sq km). That's about half the size of the United States!

← *This is a picture of the* Endeavour *as it orbited Earth during the Shuttle Radar Topography Mission.*

Working in Space

To keep track of the radar mapping of Earth, the astronauts divided into two teams. Each team took turns sleeping and working. The teams switched off every 12 hours. By Sunday morning, February 13, 2000, over 7.62 million square miles (19.7 sq km) of Earth had been mapped. The *Endeavour* flew in a low orbit so that the radar could reach Earth. The *Endeavour* astronauts carefully changed the many tapes that recorded the SRTM **data**. Things went so smoothly that the astronauts were able to skywatch out the shuttle's windows and take snapshots of Earth. Commander Kevin Kregel enjoyed feeling weightless in space. He practiced doing somersaults and often ate his meals floating up to the *Endeavour's* ceiling.

This picture shows (clockwise from left) *astronauts Janet Kavandi, Dominic Gorie, and Mamoru Mohri aboard the* Endeavour.

15

Interview with Kevin Kregel

Was there any special training necessary for this mission?
We spent a lot of time working on operating the mast, how to extend and then retract it. And Janet Kavandi and Gerhard Thiele worked on a space walk in case we had to deploy the mast manually. Luckily we didn't need to do that.

Space walks can be dangerous, isn't that true?
Oh sure. Any EVA (extravehicular activity) brings risk with it. That and liftoff are two of the most dangerous operations. With liftoff, there are always a lot of things that can go wrong.

Speaking of liftoff, is it painful?
I wouldn't say painful. We pull, at most, three times the force of gravity, which is much lighter than they used to be, say, during the time of the Apollo missions. The seats are a bit uncomfortable, though. At liftoff, I'd say it feels like you've got a couple of friends sitting on your chest.

On a day-to-day basis, how did this mission differ from others you've been on?
This mission was solely concerned with mapping Earth. We did a lot of the same things over and over. Not that anything was routine. We filled 332 tapes with data. Well, we had to change those tapes and any delay in doing so meant that we would lose data. Also, anytime we were over land we had to monitor the radar. We had to be 100 percent accurate every day of the 11-day mission. That's not easy to do.

What was the sleeping situation like?
We had four sleeping bunks with sleeping bags that were held down so that we wouldn't go floating off. They were pretty comfortable and kept much of the sound out so that we could get rest.

How do you work off stress up in space?
We had an exercise bike, which is not only good for stress, but also important for an astronaut's heart. When you are up in space, in microgravity, your heart doesn't have to work as hard as it does going through a normal day on Earth. In space, it's kind of like being on bed rest. But as you reenter Earth's atmosphere again, your heart has to start working to that same high level. If we didn't keep our hearts working hard by exercising in space, we might get dizzy as we start to land the shuttle. You certainly don't want an astronaut passing out at landing.

How's the food up there?
(laughs) Freeze-dried. We bring up a bit of fresh food, but it will go bad if it's not eaten in a day or so. The freeze-dried food has gotten much better, though. There are about 150 different kinds.

How tight are the quarters up there?
The shuttle is about the size of an RV that's been cut in half with one half stacked on top of the other. I'm used to it, but during meals, I usually float to the top of the ceiling to eat. That's not something I can do at home.

I'd say not. Well, Commander Kevin Kregel, thank you so much for taking the time to discuss this mission. It certainly was an historic one.
Yes, it was. And you're very welcome. It was my pleasure.

Communication

The *Endeavour* SRTM crew **communicated** with scientists at **Mission Control** through radio communication. People around the world wanted to know about the SRTM mission, too. Almost every day, television reporters on Earth used **satellite** communication to ask the crew about their work. Another communication system was for kids only. Students in 80 schools in the United States, France, Germany, and Japan used the Internet to communicate with a special camera, called EarthKAM. EarthKAM was mounted on a window of the shuttle. Students used school computers to ask EarthKAM to take photographs. The pictures were transferred to laptop computers and could be seen by the students in about 2 hours.

In this picture, astronauts Kevin Kregel (left) *and Gerhard Thiele* (right) *are shown on a screen talking to Mission Control from the* Endeavour.

20

Building the Map

The SRTM system on the *Endeavour* used radar to record images of small, square patches of land. Each small patch was put together with millions of other small patches. Computers were later used to blend all of the squares together into a single map. Scientists believe that the information collected by the astronauts could fill more than 20,000 CDs. By February 18, 2000, one week into the mission, more than 88 percent of the target area had been mapped. The SRTM radar system worked very quickly. It could map all of Alaska, the biggest state in the United States, in about 15 minutes. Mapping the smallest state, Rhode Island, took less than 2 seconds!

← *This is a view of the Hawaiian islands Maui (top left), Molokai (bottom left), and Lanai (right), taken during the Shuttle Radar Topography Mission.*

Coming Home

By Monday, February 21, 2000, the SRTM mapping was done. Before the *Endeavour* could land, though, the astronauts had to fold the mast into its canister. The locks on the canister didn't work because they were too cold to move. By warming up the locks and pulling harder on the mast, the crew was able to store the mast.

The *Endeavour* could have landed at either the Kennedy Space Center in Florida or at Edwards Air Force Base in California. Mission Control was worried that winds in Florida would blow the *Endeavour* off the runway as it glided down from space. Mission Control told the *Endeavour* that it was safe to land in Florida. On Tuesday, February 22, the *Endeavour* landed at the Kennedy Space Center in Florida.

Glossary

antennae (an-TEH-nee) Metal objects used to send and receive signals.
canister (KAH-nih-ster) A container.
communicated (kuh-MYOO-nih-kayt-ed) Having shared information or feelings.
data (DAY-tuh) Information.
depth (DEPTH) How deep something is.
fault lines (FAWLT LYNZ) Breaks in Earth's crust formed by an earthquake.
landslides (LAND-slydz) The fast movement of rock, earth, or human-made material down a slope.
launchpad (LAWNCH-pad) The platform from which a spacecraft is sent into space.
Mission Control (MIH-shun kun-TROHL) A group of scientists who guide space travel from the ground.
mission specialists (MIH-shun SPEH-shuh-lists) Members of a space shuttle crew who perform a certain job.
NASA (NA-suh) National Aeronautics and Space Administration, the United States's space agency.
orbits (OR-bihts) The paths that one body takes around another, usually larger, body.
payload bay (PAY-lohd BAY) The area on a space shuttle where equipment is stored.
propulsion (pruh-PUL-zhun) The force that moves something forward.
radar (RAY-dar) A system that uses sound waves to locate objects.
rigid (RIH-jid) Stiff.
satellite (SA-til-eyet) A human-made or natural object that orbits another body.
thruster rockets (THRUST-er RAH-kets) Rockets that help a spacecraft move in space.
topography (tuh-PAH-gruh-fee) A way of describing the shape of different landmasses, such as the height of mountains.

Index

E
Earth, 6, 9, 10, 13, 14, 16–18
EarthKAM, 18
earthquake(s), 5, 6
European Space Agency, 5

G
Gorie, Dominic, 5

J
Jet Propulsion Laboratory, 6

K
Kavandi, Janet, 5
Kennedy Space Center, 5, 22
Kevin, 5, 14, 16, 17

L
landmasses, 5, 6, 10

M
Mission Control, 18, 22
Mohri, Mamoru, 5

N
NASA, 6
National Imagery and Mapping Agency, 6

O
orbit(s), 6, 10, 13, 14

R
radar, 5, 9, 10, 13, 14, 16, 21

S
satellite communication, 18

T
Thiele, Gerhard, 5

V
Voss, Janice, 5

Web Sites

To find out more about the *Endeavour* SRTM, check out these Web sites:
www.earthkam.ucsd.edu
www.jpl.nasa.gov/srtm